Contents

KU-605-317

1

electronics today and yesterday

Our modern-day dependency on electronics was spurred by the invention of the transistor in the late 1940s, which has ultimately led to unbelievably complex circuits containing thousands of transistors integrated on a sliver of silicon so small you could lose it under your fingernail. This chapter begins by describing a few of the ways this 'new' technology has come to influence almost every aspect of our lives – whether it be work, rest or play. For example, electronics has transformed the music industry, communications, patient care and travel. This is followed by a brief history of the main developments in electronics over the years starting with discovery of the electron and the development of the thermionic valve. Once the transistor had replaced the valve, there was no stopping electronics becoming the world's fastest growing industry, culminating in the development of the silicon chip.

1.1 The electronic age

Electronics is the art of using electrons in devices such as transistors and silicon chips to make electricity work for us. There is no doubt that this 'art' has had far-reaching effects on nearly all aspects of life, although its influence on everyday life often remains unseen. Our modern-day dependency on electronics was spurred by the invention of the transistor in the late 1940s. This was followed by the manufacture of silicon chips in the early 1960s that has ultimately led to incredibly complex circuits containing thousands of transistors integrated on a sliver of silicon so small that you could lose it under your fingernail. Miniaturizing electronic circuits in this way is called microelectronics and it has come to influence the way we store, process and distribute information; to change the way we design and manufacture industrial goods; to improve the diagnosis and treatment of illnesses; and it shapes the affairs of finance and business, as well as a variety of social, educational and political activities. Nowadays, we take for granted the way electronics makes our lives more comfortable, enjoyable, creative and exciting, so let us begin the study of electronics by taking a closer look at some of its applications.

1.2 Communications electronics

It is difficult to ignore the ubiquitous mobile phone that has brought about such remarkable changes in people's daily lives, although when used carelessly in public places it is often associated with annoyance to others. Energized by people's need to communicate with others whether for business or pleasure, the development of the mobile phone has been rapid and widespread enabling us to keep in touch with others while on the move almost anywhere on the planet. Moreover, the demand for Internet access and for other services via the 'mobile' is stimulating a wide range of facilities for this modern personal communication device that gets progressively 'smarter'.

1.3 Computer electronics

Compared with the first computers of the 1940s, today's computers show how remarkable the advances in electronics have been. Power-hungry, room-sized and unreliable, these early computers have been replaced by a variety of compact, efficient and versatile computers such as laptops. Computers are essential for the reliable and safe operation of navigation systems and landing systems for aircraft, and for automatic motorway toll stations which provide revenue for new roads and environmentally friendly traffic flow systems. A computer loaded with the appropriate software can meet anyone's needs: the tourist wishing to create a video of a holiday in the Caribbean; an astronomer interested in studying light from a distant galaxy; an engineer planning a new motorway; an accountant working out the cost of running a business; all these and more make the computer a powerful tool.

1.4 Control electronics

Electronics is an essential part of modern control systems. For example, anti-skid braking systems (ABS) on some cars depend on electronic devices that ensure that wheels do not lock and cause a skid. Safety in passenger and military aircraft relies on complex control systems. Moreover, the regulation of temperature and pressure in chemical factories and nuclear power stations depends on electronic control systems for the safe and efficient operation of the plant. In the home, the power of electric drills and food mixers can be controlled at the turn of a knob. In the greenhouse, automatic control of temperature and humidity is possible. Under the control of a program, the microprocessor in a washing machine instructs devices such as motors, heaters and valves to carry out a prearranged sequence of washing processes.

1.5 Medical electronics

Electronic equipment is widely used to diagnose the cause of illnesses, to help nurses care for patients and to be of assistance to surgeons carrying out operations. For example, the electrocardiograph (ECG) is an electronic instrument that can help in diagnosing heart defects. By means of electrodes attached to the body, the ECG picks up and analyses the electrical signals generated by the heart. The ECG is just one of a number of instruments available to nurses and doctors in hospitals. A complete patient-monitoring system also records body temperature, blood pressure and heart rate and warns staff should there be any critical change.

1.6 Looking back

More than a century ago, electronics was unknown. There were no radios, televisions, computers, robots or artificial satellites, and none of the products and services described above that we now take for granted. This revolution came naturally out of a study of electricity, familiar to the Greeks over 2000 years ago. Electricity was a subject of great interest to Victorian scientists, and to Sir William Crookes in particular.

1.7 The discovery of cathode rays

The beginnings of electronics can be traced back to the discovery of cathode rays in the closing years of the nineteenth century. These mysterious rays had been seen when an electrical discharge took place between two electrodes in a glass tube from which most of the air had been removed. Sir William

Crookes called these 'cathode rays' since they seemed to start at the negative electrode (the cathode) and moved towards the positive electrode (the anode).

At that time, nobody had any idea what cathode rays really were. Nevertheless, during a historic lecture at the Royal Institution in London in April 1897, Sir J. J. Thomson declared that cathode rays were actually small, rapidly moving electrical charges. Later these charges were called electrons after the Greek word for amber.

1.8 The invention of the thermionic valve

The first practical application of cathode rays was the invention of the thermionic valve by Sir John A. Fleming in 1904. In this device, the heating of a wire (the filament) in an evacuated glass bulb produces electrons. The word 'thermionic' comes from 'therm' meaning 'heat', and 'ion' meaning 'charged particle', i.e. the electron. In a valve, negatively charged electrons driven out from the heated filament (the cathode) moved rapidly to a more positive anode. The flow of electrons stops if the anode becomes more negative than the cathode. This electronic component is called a diode since it has two (so 'di') electrodes for making connections to an external circuit. In addition, it acts like a valve because electrons flow through it only in one direction, from the cathode to the anode, not in the opposite direction.

It did not take long for an American, Lee de Forest, to make a much more interesting and useful thermionic valve. By adding a third electrode, made of a mesh of fine wire through which the electrons could pass, he produced a triode. By adjusting the voltage on this third electrode (called the grid), he was able to make the triode behave like a switch and, more importantly, as an amplifier of weak signals. The triode made it possible to communicate over long distances by radio.

1.9 The beginnings of radio and television

Strangely, the First World War (1914–18) did little to stimulate applications for thermionic valves. But immediately after the war, electronics received a push that has gained strength ever since. In London the British Broadcasting Corporation was formed, and in 1922 its transmitter (call sign 2LO) went on the air.

The second major boost to the emerging electronics industry was the start of regular television transmissions from Alexandra Palace in London in 1936. But at that time the public had very little interest in television, which was hardly surprising, as the pictures produced by John Logie Baird's mechanical scanning system were not very clear. By the time an electronic scanning system had been developed that gave much better pictures, the Second World War had begun. The Alexandra Palace transmitter was closed down abruptly in September 1939 at the end of a Mickey Mouse film for fear that Germany might use the transmission as a homing beacon for its aircraft to bomb London.

1.10 The invention of the transistor

In the period immediately following the Second World War, there was a major step forward in electronics brought about by the invention of the first working transistor. In 1948, Shockley, Barden and Brattain, working in the Bell Telephone Laboratories in the USA, demonstrated that a transistor could amplify electrical signals and also act like a switch. However, the way electricity moved in semiconductors, as these germanium-based devices were called, was not well understood. Furthermore, until the 1950s it was not possible to produce germanium (and later silicon) with the high purity required to make useful transistors.

These transistors turned out to be successful rivals to the thermionic valve. They were cheaper to make since their manufacture could be automated. They were smaller, more rugged and had a longer life than valves, and they required less electrical power to work them. Once silicon began to replace germanium as the basic semiconductor for making transistors in the 1960s, it was clear that the valve could never compete with the transistor for reliability, compactness and low power consumption.

1.11 Silicon chips make an impact

The first integrated circuits were made during the early 1960s. Techniques were developed for forming up to a few hundred transistors on a silicon chip and linking them together to produce a working circuit. The *Apollo* spacecraft that took men to the Moon in the late 1960s and early 1970s used these third-generation computers for navigation and control. The stimulus to miniaturize circuits in the form of integrated circuits came from three main areas: weapons technology, the 'space race' and commercial activity.

The 1970s saw the number of transistors integrated on a silicon chip doubling every couple of years and this trend continues. Along with this increasing circuit complexity has been a similar doubling in the information processing power of the silicon chip.

2

simple circuits and switches

Electronics is dependent on movement of electrons – that is, an electric current. But what makes electrons move? An electric force of course! This chapter begins by looking into this important idea of electron flow. Electronic components such as transistors, resistors and capacitors are commonly arranged in series (one after the other) and/or in parallel (side by side). This chapter explains this distinction, then goes on to develop an understanding of the relationship between current, voltage and resistance. Resistance is an important electronics concept for it determines the distribution of current and voltage in a circuit. We next turn our attention to switches, first looking at mechanical switches, which are common to many household appliances, for example, although increasingly they are being replaced by solid state switches based on transistors. Next, we see how simple switches are used in decision-making digital circuits.

2.1 Making electrons move

Just as a gravitational force is required to make water flow downstream, an electrical force is required to make an electron move through a conductor. If the conductor is a copper wire as shown in Figure 2.1, electrons move through it when there is a difference of electrical force between its ends. This force is called a potential difference (p.d.) and is measured in volts (symbol V). Since a lamp lights when it is connected across the terminals of a battery, there must be a potential difference between these terminals to make current flow through the lamp. The electrical force, or potential difference, provided by a battery is known as an electromotive force (e.m.f.) (symbol E) of the battery and this is also measured in volts.

The flow of electrons through the lamp shown in Figure 2.2 is called an electric current (symbol I) and is measured in amperes (symbol A). A great many electrons are on the move in a conductor when a current of one ampere flows through it. In fact, about six million million million move past a point in the circuit each second! Since each electron carries an electrical charge, this current is a flow of electrical charge through the circuit. Electrical charge (symbol Q) is measured in coulombs (symbol C). When a current of one ampere flows through a circuit, the rate at which charge flows is one coulomb per second. Thus:

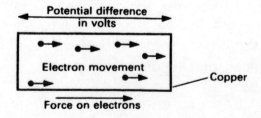

Figure 2.1 *The flow of electrons through copper.*

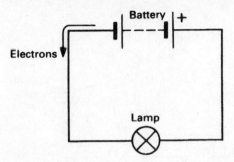

Figure 2.2 *A simple circuit.*

> 1 ampere = 1 coulomb per second
> or 1 A = 1 C/s
> or, in terms of quantities, $I = Q/t$

2.2 Series and parallel circuits

Devices such as lamps, switches, batteries and transistors are known as components. They are the individual items that are connected together to make a useful circuit. Figure 2.3 shows a simple circuit in which a battery, B_1, that has an e.m.f. of 6 V makes a current I flow through lamps L_1 and L_2. The switch SW_1 has two positions, open and closed. When the switch is closed, it offers a low resistance path and allows current to flow round the circuit. If SW_1 is open, it offers a high resistance path (the resistance of the air between the switch contacts) and stops current flowing through the circuit. The current comprises electrons that flow from the negative to the positive terminal of the battery. Before electrons were discovered, it was thought that electrical current flowed from the positive to the negative terminal of a battery. This direction is known as conventional current and is usually marked by an arrow on circuit diagrams. The circuit of Figure 2.3 is called a series circuit since the two lamps, the battery and the switch are connected one after the other. In a series circuit, the current

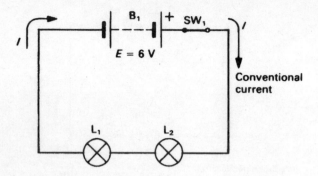

Figure 2.3 *Two lamps connected in series.*

flowing is the same at any point in the circuit so that the same current flows through each lamp.

Another common circuit arrangement of components that you will meet in your study of electronics is shown in Figure 2.4. In this circuit, two identical lamps, L_1 and L_2, are connected side-by-side to a battery, B_1, of e.m.f. six volts (6 V). Each lamp has the same p.d. across it, i.e. 6 V, since each is connected across the battery. When both switches are closed, the current flowing through each lamp is 0.06 A. So the total current provided by the battery is 0.12 A, the sum of the currents flowing through the two lamps. The circuit of Figure 2.4 is known as a parallel circuit.

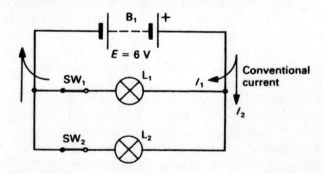

Figure 2.4 *Two lamps connected in parallel.*

electronics made easy

In this circuit, switch SW_1 independently controls lamp L_1, and SW_2 independently controls L_2. In the home, different appliances are connected in parallel with the mains supply so that each one is capable of being controlled by their respective on/off switches.

2.3 Resistance and Ohm's law

Electrical resistance (symbol R) is a measure of the ease (or rather the difficulty!) with which electrical current is able to flow through a material. Copper has a low resistance since it is a good conductor of electricity; glass has a high resistance since it is a very poor conductor, i.e. an insulator. The electrical resistance of a material is measured in units of ohms and is defined by the following equation:

$$\text{resistance} = \frac{\text{p.d. across the material}}{\text{current through the material}}$$

$$\text{or } R = \frac{V}{I}$$

This equation can be used to find the resistance of one of the filament lamps shown in Figure 2.4. The p.d. across each lamp is 6 V, and is the same as the e.m.f. of the battery. The current through each lamp is 0.06 A. Thus, the resistance, R, of each lamp is found as follows:

$$R = \frac{6\text{ V}}{0.06\text{ A}} = 100 \text{ ohms}$$

The unit of electrical resistance is the ohm; the symbol given to it is the Greek letter, Ω (omega). So the lamp has a resistance of 100 Ω. Of course, we could use the equation to find the current through a component, or the p.d. across a component. We would then need to use the above equation in a different form. Figure 2.5 helps to get the equations right.

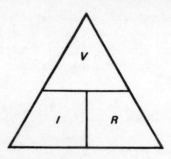

Figure 2.5 *A triangle for working out V, I and R.*

To find R : cover R and $R = V/I$

To find V : cover V and $V = I \times R$

To find I : cover I and $I = V/R$

For example, suppose you want to find the current flowing through a 12 V car headlamp bulb that has a resistance of 3 Ω. Since the current $I = V/R$,

$$I = \frac{12\,V}{3\,\Omega} = 4\ A$$

Note: The symbol, Ω, will be omitted from resistor values in all of the circuits in this book. Instead resistor values will be indicated by the multiplier 'R' for resistor values less than 999 ohms, by 'K' for resistor values between 1000 ohms and 99,999 ohms, and by 'M' for resistor values above and including 1,000,000 ohms. However, in the text, for clarity, the ohms symbol will be used.

If the resistance of a component is constant for a range of different values of potential difference and current, the component is said to be linear, or ohmic, and it obeys Ohm's law. Ohm's law is as follows:

Provided the temperature and other physical conditions of an electrical conductor remain unchanged, the potential difference across it is proportional to the current flowing through it.

2.4 Large and small numbers

In electronic circuits, it is usual for values of current and potential difference to be small while values of resistance are large. For example, the value of a resistance might be one million ohms (1,000,000 Ω). Instead of writing down all the zeros for this large number, it is much easier to use the prefix 'M' meaning 'mega' for one million and write the resistance as 1 MΩ (or just 1 M on circuit diagrams). Similarly, small values of current can be expressed using the prefix 'm' meaning 'milli' for 'one thousandth of', or 'μ' meaning 'micro' for 'one millionth of'. Thus, ten milliamperes can be written as 10 mA, and 100 micro amperes as 100 μA. The table below summarizes the values of some of the prefixes you can expect to meet in working with electronic circuits.

Table 2.1   Large and small numbers.

Prefix	Factor	Powers of ten	Symbol
Tera	1 000 000 000 000	10^{12}	T
Giga	1 000 000 000	10^{9}	G
Mega	1 000 000	10^{6}	M
Kilo	1000	10^{3}	k
Milli	0.001	10^{-3}	m
Micro	0.000 001	10^{-6}	μ
Nano	0.000 000 001	10^{-9}	n
Pico	0.000 000 000 001	10^{-12}	p

When working with quantities in electronics that have large and small values, you will find it helpful to express the values as powers of ten rather than by writing down many zeros. The powers of ten equivalent to the prefixes are also listed in the table. Thus, the factor 1000 meaning 'one thousand times' is expressed as 10^3, meaning 'ten to the power three' ($10 \times 10 \times 10 = 1000$). In addition, the factor 0.000 001 meaning 'one millionth of' is expressed as 10^{-6}, meaning 'ten to the power minus 6' ($1/(10 \times 10 \times 10 \times 10 \times 10 \times 10)$). The numbers above the tens are called indices. Calculations become easier when large and small numbers are expressed as powers of ten because the indices can be added or subtracted.

2.5 Types of switch

Switches are used to turn current fully on or fully off in a circuit, not to some in-between value of current. A switch is 'on', or closed, when it offers a low resistance path for current flowing through it, and it is 'off', or open, if it offers a high resistance path to current flowing through it. We operate simple on/off switches dozens of times a day. A car's ignition, heater and radio are operated using on/off switches. So are cookers, hi-fi systems, televisions, radios and burglar alarms in the home. Moreover, the keyboard or keypad switches of calculators, computers and electronic games act to switch something on and off.

Many switches in everyday use require a mechanical force to operate them. The force brings together or separates electrically conducting metal contacts. Three types of mechanical switch are shown in Figure 2.6. The push-button switch (Figure 2.6a) is a simple push-to-make, release-to-break type. There are two circuit symbols for this type of switch depending on whether pushing makes or breaks the contacts. Slide and toggle switches (Figure 2.6b) are generally made either as single-pole, double-throw (s.p.d.t.), or as double-pole, double-throw (d.p.d.t). The poles of these switches are the number of separate circuits that the switch will make or break

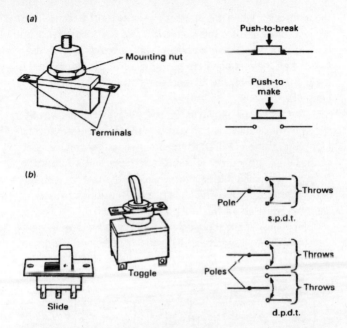

(a)

Mounting nut

Terminals

Push-to-break

Push-to-make

(b)

Slide

Toggle

Pole

Throws

s.p.d.t.

Poles

Throws

Throws

d.p.d.t.

Figure 2.6 *(a) Push-to-make, release-to-break switch; (b) Slide and toggle switches.*

simultaneously. Thus, a d.p.d.t switch can operate two separate circuits at the same time. An s.p.d.t switch is sometimes known as a change-over switch, since the pair of contacts that is made changes over as the switch is operated.

2.6 Simple digital circuits

Switches like the ones described above are on/off components. There is no in-between state enabling them to be in 'half on or half off', or 'nearly on or nearly off'. To explain the significance of this rather obvious statement, look at Figure 2.7 showing a simple circuit comprising a switch SW_1 connected in series with a battery and a lamp, L_1. When the switch is closed, the lamp is on; when it is opened, the lamp is off. Because there are just two states for this circuit, it is said to be a digital circuit. If we use the two binary numbers, 1 and 0, to represent these two states, the number 1 represents the switch closed and the lamp on, and the number 0 represents the switch open and the lamp off.

The 'black box' drawn in Figure 2.7b is a symbolic way of showing the circuit. The two states of the switch represent the input information (1 or 0) to the box. In addition, the two states of the lamp, either on or off, represent the output information from the box. The table in Figure 2.7c summarizes the output and input information. The table is called a truth table since it 'tells the truth' about the function of the circuit.

Now look at the more complicated circuit shown in Figure 2.8a in which two switches, SW_1 and SW_2, are connected in

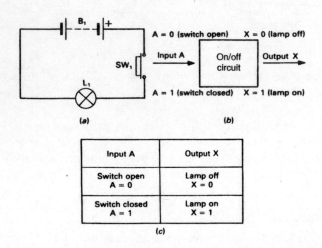

Input A	Output X
Switch open A = 0	Lamp off X = 0
Switch closed A = 1	Lamp on X = 1

(c)

Figure 2.7 (a) A simple on/off circuit; (b) the circuit shown as a functional black box; (c) the truth table for this on/off circuit.

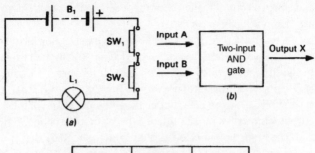

Figure 2.8 *(a) A simple series circuit; (b) the circuit shown as a functional black box; (c) the truth table for this two-input AND gate.*

Input A (SW$_1$)	Input B (SW$_2$)	Output X
0	0	0
1	0	0
0	1	0
1	1	1

(c)

series. Note that the lamp cannot light unless SW$_1$ and SW$_2$ are closed. As the black box in Figure 2.8b shows, the switches provide input information and the lamp indicates the output information. The truth table in Figure 2.8c summarizes the function of this digital circuit known as an AND gate. It 'says' that the output state has a binary value of 1 (lamp on) only if switch SW$_1$ and switch SW$_2$ each have a value of binary 1 (both switches closed). If either or both of the switches are set to binary 0 (are open), the output is binary 0 (the lamp is off). Note that this digital circuit is called a gate because the switches open and close to control the information reaching the output just as a gate can be open or closed.

A second simple digital circuit is shown in Figure 2.9a in which the two switches are connected in parallel. In this circuit the lamp lights if switch SW_1 or switch SW_2 is closed. The lamp also lights if both switches are closed. This OR gate has the truth table shown in Figure 2.9b that summarizes the values of the output information for all values of the input information.

The branch of electronics that is introduced by the above series and parallel circuits is known as digital logic. The AND and OR gates are called logic gates since their output states are the logical (i.e. predictable) result of a certain

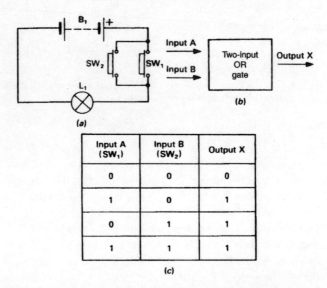

Input A (SW_1)	Input B (SW_2)	Output X
0	0	0
1	0	1
0	1	1
1	1	1

(c)

Figure 2.9 (a) A simple parallel circuit; (b) the circuit shown as a functional black box; (c) the truth table for this two-input OR gate.

combination of input states. These logic gates, and others besides, are of great importance to electronics nowadays. Many thousands of logic gates are built from transistors in the form of integrated circuits that are used in a variety of equipment from control and communications equipment to calculators, watches and computers.

3

voltage dividers and resistors

The explanation of how a voltage divider controls voltage requires a little bit of mathematics and an understanding of the relationship between current, voltage and resistance. We also note the effective resistance of resistors connected in series and parallel, requiring a little more maths. Resistors control voltage; and voltage is what makes circuits work. Nowadays, resistors are small so instead of writing the value on them, they are colour coded using at least three coloured bands to make them easy to see, so you can practise reading the colours. Finally, there are two components that are very useful in electronics. The first one is the light dependent resistor, the resistance of which changes with changes in light intensity. They are used in automatic streetlights, for example. The second is the thermistor, the resistance of which changes with temperature, so it is used in such devices as fire alarms and thermostats.

3.1 The potential divider

Figure 3.1a shows the function of this useful circuit building block. It provides an output potential difference, V_{out}, that is less than the input potential difference, V_{in}. However, why should this be a useful function? Well, potential dividers are used as volume controls in radios, and for controlling the brightness of television screens, for example. As you will see, potential dividers are widely used in other circuit designs, e.g. in control and instrumentation systems where a voltage has to be reduced to a value suitable for operating transistors and integrated circuits, or for fixing a voltage at a preset level.

Figure 3.1b shows a circuit that acts as a potential divider by dividing the e.m.f. generated by a battery. The two rectangular symbols, marked R_1 and R_2, are electronic components called resistors; they simply have an electrical resistance measured in ohms. The reduced output voltage measured with respect to 0 V, occurs at the junction between the two resistors. It is possible to obtain any value of p.d. between 0 V and V_{in}, the e.m.f. of the battery, by changing the

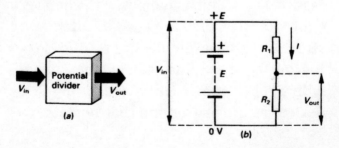

Figure 3.1 (a) The function of a potential divider; (b) The use of two resistors as a potential divider.

values of the resistors. The values of the resistors R_1 and R_2 determine the output p.d. V_{out}. The equation is

$$V_{out} = \frac{V_{in} \times R_2}{R_1 + R_2}$$

This equation shows that V_{out} is less than V_{in} by the fraction $R_2/(R_1 + R_2)$, i.e. the smaller R_2, the smaller V_{out}. Just suppose that $V_{in} = 9\,V$ and the values of the resistors are $R_1 = 90\,\Omega$ and $R_2 = 10\,\Omega$. Now

$$V_{out} = \frac{9\,V \times 10}{100} = 0.9\,V$$

The same value for V_{out} could have been obtained if $R_1 = 900\,\Omega$ and $R_2 - 100\,\Omega$, for then

$$V_{out} = \frac{9\,V \times 100}{1000} = 0.9\,V$$

Or if $R_1 = 240\,\Omega$ and $R_2 = 120\,\Omega$, then

$$V_{out} = \frac{9\,V \times 120}{360} = 3\,V.$$

Note that it is the ratio of the values, not the actual values of the resistors, that determines the output voltage of a potential divider. You can prove the above equation very simply by using the relationship between V, I and R that was given in Chapter 2. First note that the current, I, flows through both resistors, and is given by

$$I = \frac{V_{in}}{R_1 + R_2}$$

In addition, V_{out} is given by $I \times R_2$.

If we substitute I from the first equation into the second,

$$V_{out} = \frac{V_{in} \times R_2}{R_1 + R_2}$$

as required.

3.2 Resistors in series and parallel

The combined resistance of two resistors connected in series is found by adding together their values. Thus in Figure 3.2a the total resistance, R, of two resistors R_1 and R_2 connected in series is given by the equation

$$R = R_1 + R_2$$

This equation can be easily proved. First, note that when two resistors are connected in series the same current, I, flows through each resistor. Second, the sum of the p.d.s across the two resistors is equal to the p.d. across the combination. Thus $V = V_1 + V_2$. And since $V = I \times R$, we can write

$$V = I \times R_1 + I \times R_2 = I \times (R_1 + R_2) = I \times R$$

Here we have written $R = R_1 + R_2$ for the resistance of the combination. Thus, if we replace the two resistors connected in series by a single resistor equal to the sum of the combination, the current drawn from the battery remains unaltered.

As shown in Figure 3.2b, when two resistors are connected in parallel the total resistance of the combination is given by the equation

$$\frac{1}{R} = \frac{1}{R_1} + \frac{1}{R_2}$$

This can be rewritten

$$R = \frac{R_1 \times R_2}{R_1 + R_2}$$

This equation can be proved by first noting that the p.d. across each resistor is equal to the p.d. across the combination. Second, the sum of the currents through each resistor is equal to the current flowing from the power supply. Thus $I = I_1 + I_2$. And since $I = V/R$, and $I_1 = V/R_1$ and $I_2 = V/R_2$, we can write

$$I = \frac{V}{R} = \frac{V}{R_1} + \frac{V}{R_2}$$

This equation reduces to

$$\frac{1}{R} = \frac{1}{R_1} + \frac{1}{R_2}$$

Note that when two or more resistors are connected in series, their total resistance is more than the largest value present. On the other hand, when two or more resistors are connected in parallel, their total resistance is less than the smallest value present. We can prove this by taking two values for R_1 and R_2, e.g. let $R_1 = 300\ \Omega$ and $R_2 = 500\ \Omega$. If these two resistors are connected in series, their combined resistance is $300 + 500 = 800\ \Omega$, that is more than the largest value present, i.e. more than $500\ \Omega$. Additionally, if they are connected in parallel, their combined resistance is

$$\frac{300 \times 500}{300 + 500} = \frac{150000}{800} = 187.5\ \Omega$$

This is less than the smallest value present, i.e. less than $300\ \Omega$.

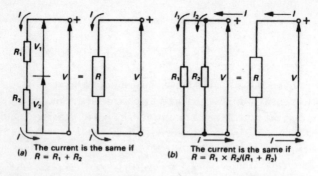

Figure 3.2 (a) Two resistors connected in series; and (b) two resistors connected in parallel.

3.3 Fixed-value and variable resistors

Figure 3.3 shows examples of fixed-value resistors. By depositing a hard crystalline carbon film on the outside of a ceramic rod, a carbon-film resistor is made. It is then protected by means of a hardwearing and electrically insulating coating. The resistance of the carbon film between the connecting wires is the resistor's value. Of similar construction is the metal-film resistor, except that tin oxide replaces the carbon. The metal-film resistor has better temperature stability than the carbon-film resistor. Both types are recommended for use in audio amplifiers and radio receivers where they may be exposed to extreme changes of temperature and humidity. Metal-film resistors also generate little electrical 'noise'. (Electrical noise is the 'hiss' that can tend to 'drown' the required signal in a circuit, and is caused by the random movement of electrons in the resistor.) By winding a fine wire of nichrome (an alloy of nickel and chromium) round a ceramic rod, a wire-wound resistor

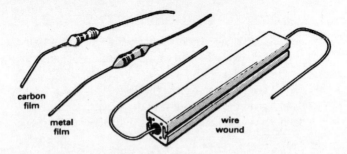

Figure 3.3 *Examples of fixed-value resistors.*

is made. Wire-wound resistors can be made to have a very precise value, guaranteed to within 0.1 per cent.

The thick-film resistor is made by adjusting the thickness of a layer of semiconducting material to give the required resistance. These resistors are generally grouped eight at a time in a single-in-line or dual-in-line package (Figure 3.4). The individual resistors may be independent of each other or have a common connection, depending on the application. Thick-film resistors of this type are particularly useful in computer circuits where eight or more connections have to be made between the computer and display or control circuits.

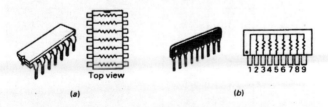

Figure 3.4 *Resistors grouped together as (a) a dual-in-line (d.i.l.) package; and (b) a single-in-line (s.i.l.) package.*

3.4 Values and coding of resistors

Most fixed-value resistors are marked with coloured bands so that their values can be read easily. Four coloured bands are used as shown in Figure 3.5.

Band 1 gives the first digit of the value.

Band 2 gives the second digit of the value.

Band 3 gives the number of zeros that follow the first two digits.

Band 4 gives the 'tolerance' of the value worked out from the first three bands.

For example, suppose the bands are coloured as follows:

Band	1	2	3	4
Colour	yellow	violet	red	silver
Value	4	7	00	10%

This resistor has a value of 4700 Ω to within 10 per cent more or less. That is, its value is 4.7 kΩ ± 10%. If the value of this resistor were measured accurately, its resistance should

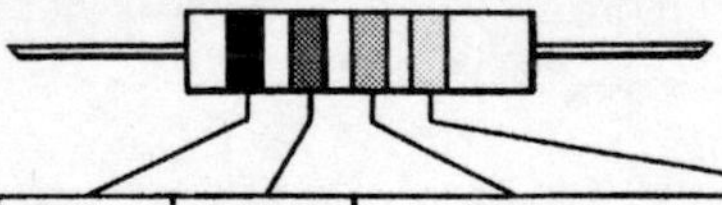

Colour	Band 1	Band 2	Band 3	Band 4
Black	0	0	None	
Brown	1	1	0	1%
Red	2	2	00	2%
Orange	3	3	000	3%
Yellow	4	4	0 000	4%
Green	5	5	00 000	–
Blue	6	6	000 000	–
Violet	7	7	0 000 000	–
Grey	8	8	–	–
White	9	9	–	–
Gold	–	–	0·1	5%
Silver	–	–	0·01	10%
No colour	–	–	–	20%

Figure 3.5 *The resistor colour code.*

electronics made easy

not be more than 4.7 kΩ ± 0.47 kΩ, or less than 4.7 kΩ – 0.47 kΩ, i.e. between 5.17 kΩ and 4.23 kΩ. For most circuit designs, it is unnecessary to use resistors with tolerance less than 5 per cent.

Instead of using a colour code, some manufacturers are marking the values of resistors using the British Standards BS1852 code. This code is often used to mark resistor values on circuit diagrams, as in this book. The BS1852 code consists of letters and numbers as the following examples show.

BS1852 code	Resistance
6K8M	6.8 kΩ± 20%
R47K	0.47 Ω± 10%
5R6J	5.6 Ω± 5%
47KG	47 kΩ± 2%
2M2F	2.2 MΩ± 1%

Note that in the BS1852 code, instead of the decimal point a letter, e.g. 'K', is used to indicate the multiplying factor. Thus in the code 6K8M, the K indicates that the resistor has a value of 6.8 × 1000 Ω – 6.8 kΩ.

3.5 Special types of resistor

Some resistors change their resistance in response to the change in some property. Two of the most useful devices of this type are shown in Figure 3.6. The resistance of the light dependent resistor (LDR) changes with the amount of light falling on it. The resistance of the thermistor changes with its temperature. The graphs show how an ohmmeter records the change in resistance of these devices. The resistance of the LDR increases greatly from daylight to darkness, while the resistance of the thermistor would increase by a much smaller amount if its temperature fell from, say, 100°C to 0°C. Both the LDR

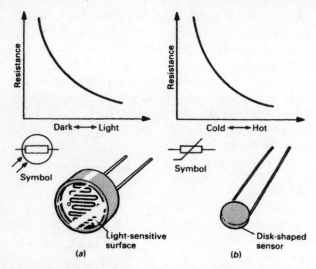

Figure 3.6 *Two special types of resistor: (a) the light-dependent resistor; and (b) the thermistor.*

and the thermistor are based on semiconductors. The LDR uses a material such as cadmium sulphide, and the thermistor a mixture of different semiconductors.

Light-dependent resistors are used in photographic light meters, security alarms and automatic street light controllers. Thermistors are used in control systems such as thermostats, in fire alarms and in thermometers.

The LDR and thermistor are generally used as one of the resistors in a potential divider as shown in Figure 3.7. When their resistance changes, there is a change of p.d. across them. Thus, if we use the following relationship (Section 5.1)

$$V_{\text{out}} = V_{\text{in}} \times \frac{R_2}{R_1 + R_2}$$

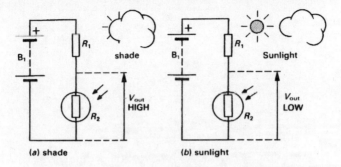

Figure 3.7 *Using the LDR in a potential divider.*

In this equation, R_2 is the resistance of the LDR or thermistor. It is easy to see what happens when the resistance of the LDR or thermistor changes. Figure 3.7a shows that when the LDR is in shade, its resistance, R_2, is high so that the p.d. across it (i.e. V_{out}) is high. If it is in sunlight, its resistance falls, so V_{out} falls.

Similarly, if the thermistor replaces the LDR in this potential divider, high temperature makes the resistance, R_2, of the thermistor low and so V_{out} is low. A low temperature increases the thermistor's resistance so that V_{out} is high. Note that there is a smooth change of output voltage from the potential divider with change of light intensity or temperature. Thus, the potential divider is an analogue device, since V_{out} changes smoothly with change of temperature or light intensity. The potential divider in which one resistor is an LDR or a thermistor is a very useful arrangement in control and instrumentation circuits, usually forming part of a Wheatstone bridge circuit.

4

capacitors, timers and oscillators

This chapter provides a little background to two common functions of electronics, timing and oscillating, and an important component in their circuits – the capacitor. A capacitor is simply a component that stores and releases electrical charge and we give an explanation of how this happens. Unlike most resistors, which generally have a tubular shape, capacitors come in variety of shapes and this chapter describes some of these. Like resistors, capacitors may be connected in series or parallel to provide different values in total and we describe the maths required to achieve this. Capacitors normally appear in circuits connected to resistors – the combination providing a useful concept called 'time constant' – and we use graphs to show what this means. One of the components used in the design of timers and oscillators is an integrated circuit called a '555 timer', and we explain how it produces these functions.

4.1 What timers and oscillators do

Figure 4.1 shows the functions of two black boxes. First the timer: when this electronic device receives a 'trigger' signal at its input, the voltage at its output rises sharply, becoming HIGH. The output voltage remains HIGH for a time delay of T seconds, and then falls to 0 V again, the output voltage is then LOW. After the output voltage has fallen to 0 V, the timer needs another trigger signal at its input to repeat the time delay. The time delay is determined by the values of components within the timer black box. Timers are used a lot in electronic systems. Toasters, security lights, washing machines, digital clocks and watches, cameras and industrial processes use timers to ensure that a particular operation takes place for prearranged time periods.

Figure 4.2 shows the function of one type of oscillator. When this black box is switched on, its output voltage goes HIGH/LOW continually. The waveform of the signals produced by this oscillator is generally known as a rectangular wave. It is a square wave if the times for which the output voltage is HIGH and LOW are equal. Oscillators that produce these waveforms are used in alarm systems for flashing a lamp on and off, or for sounding an audio alarm from a loudspeaker. They are also used in electronic musical instruments, digital clocks and watches and

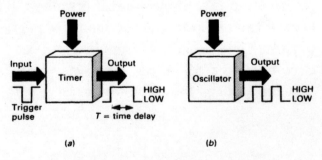

Figure 4.1 *The function of (a) a timer; and (b) an oscillator.*

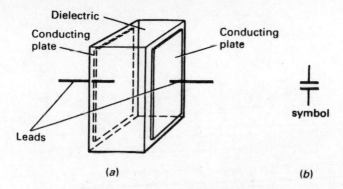

Figure 4.2 *(a) The basic structure of a capacitor; and (b) Its circuit symbol.*

in computers. The function of timers and oscillators is largely
determined by the properties of an electronic component called
a capacitor.

4.2 The way a capacitor works

Figure 4.2a shows the basic structure of a capacitor, and
Figure 4.2b the circuit symbol that reflects this structure. It
comprises two metal electrodes separated by an electrical
insulator called a dielectric. The metal electrodes are connected
to the terminals of the capacitor. The capacitor is able to store
electric charge. If a voltage is applied across the terminals of
the capacitor by a battery, B_1, as shown in Figure 4.3a, there is a
short flow of electrons in the external circuit from one electrode
to the other. Thus, one electrode becomes negatively charged
and the other positively charged. The p.d. across the terminals
is then equal to the e.m.f., E, of the battery and the capacitor
is said to be charged. That is, the excess of electrons on one
electrode, and the deficiency of an equal number of electrons on
the other electrode, represent a store of charge. If the battery is
removed, these charges remain in place since they are separated
by the dielectric which is an insulator. Now if the electrodes of

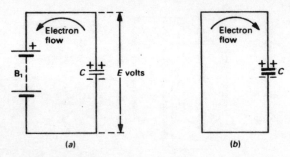

Figure 4.3 *The (a) charging; and (b) discharging of a capacitor.*

the capacitor are joined together by a conductor as shown in Figure 4.3b, electrons flow in the reverse direction through the conductor until the p.d. across the capacitor falls to zero as the charges are neutralized.

4.3 Units of capacitance

The unit of capacitance is the farad (symbol F). The farad is defined as follows: it is the capacitance of a capacitor that stores a charge of 1 coulomb when it has a p.d. of 1 V across its terminals. In general, if a charge of Q coulombs is given to a capacitor of C farads and the resulting rise in p.d. is V volts, then

$$Q = C \times V$$

The farad happens to be too large a unit to express the values of capacitors generally used in electronics, and it is necessary to use fractions of a farad as follows:

Fraction	Abbreviation
microfarad	μF (10^{-6} F)
nanofarad	nF (10^{-9} F)
picofarad	pF (10^{-12} F)

It is useful to remember that 1000 pF = 1 nF, and 1000 nF = 1 µF. Capacitors in common use range from values as small as, say, 5 pF to as large as 10,000 µF. Suppose a capacitor of value 100 µF is used in a circuit where the p.d. across it is 15 V. What is the charge stored by the capacitor? Using the equation $Q = C \times V$ above:

$$Q = 100 \times 10^{-6} \text{ F} \times 15 \text{ V}$$
$$= 1.5 \times 10^{-6} \text{ coulombs}$$
$$= 1.5 \text{ millicoulombs}$$
$$= 1.5 \text{ mC}$$

Note that this is a very small charge but it does represent the movement from one plate to the other of the capacitor of a very large number of electrons: 9.375 thousand million million to be precise! In electronics, we are not very concerned with the precise amount of charge a capacitor stores for a given voltage. We are more interested in how the charge storage properties of a capacitor are used in the design of electronic timers and oscillators.

Note: The symbol, F, will be omitted from capacitor values in all of the circuits in this book. Instead capacitor values will be indicated by the multiplier 'p' when referring to values of 10^{-12} F, by the multiplier 'n' when referring to values of 10^{-9} F, and the multiplier 'µ' when referring to values of 10^{-6} F. However, for clarity in the text, the symbol F will be included.

4.4 Types of capacitor

The value of a capacitor is usually printed on it, or it is marked with a set of coloured bands – see below. Also marked on the capacitor is its maximum safe working voltage (e.g. 100 V). If this voltage is exceeded, a surge of current through the insulating dielectric of the capacitor would certainly damage it. There are many different types of capacitor according to the

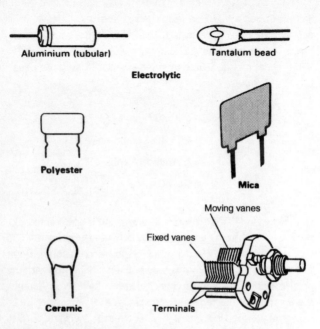

Figure 4.4 *Five types of capacitor.*

type of dielectric used in their construction. Figure 4.4 shows five different types of capacitor in general use.

Electrolytic capacitors generally have values in the range 1 μF to 50,000 μF with working voltages up to 100 V, though 450 V ones are available for thermionic valves circuits. They generally have a 'Swiss roll' construction in which the dielectric is a very thin layer of metal oxide between electrodes of aluminium or tantalum foil. The foil is rolled up to obtain a larger area of electrodes in a small volume. A tantalum capacitor offers a high capacitance in a small volume.

Polyester capacitors are examples of plastic-film capacitors. Polypropylene, polycarbonate and polystyrene capacitors are also types of plastic-film capacitor. The plastic, with the exception of polystyrene that has a low melting point, has a metallic film deposited on it by a vacuum evaporation process – these are called metallized capacitors. Metallized polyester film capacitors can have values up to 10 μF, but other plastic-film capacitors have lower values. Working voltages of polyester capacitors can be as high as 400 V, and of polycarbonate capacitors, 1000 V.

Mica capacitors are rather more expensive than plastic-film capacitors and they are made by depositing a thin layer of silver on each side of a thin sheet of mica. Like ceramic capacitors consisting of a silver-plated ceramic tube or disc, and polystyrene capacitors, they are excellent for use at high frequency. Mica capacitors have values in the range 1 pF to 10 nF, have excellent stability and are accurate to +1% of the marked value.

Variable capacitors have low maximum values, e.g. 500 pF, and their construction involves one set of metal plates moving relative to a fixed set of metal plates. The plates are separated by air or a plastic sheet that acts as the dielectric. Variable capacitors are often used in radio receivers for tuning in to different stations.

4.5 Capacitors in parallel and series

The capacitors described above fall into two main categories: polarized and non-polarized. Aluminium and tantalum capacitors, that are normally electrolytic capacitors, are polarized (see above), while polyester, polystyrene and ceramic capacitors are non-polarized. The symbols for these two types of capacitor are shown in Figure 4.5. The '+' sign on one of the terminals of the polarized capacitor indicates that this capacitor must be connected the right way round in a d.c. circuit, this leading to the positive supply or signal voltage. Note that the two parallel lines of the symbol separated by a space (filled with a dielectric) are indicative of the construction of a capacitor.

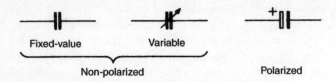

Figure 4.5 *Three symbols for capacitors.*

If the area of each plate of a capacitor is doubled, their separation remaining the same, the capacitance of the capacitor is doubled since the area is able to store twice the charge. In addition, if the separation of the plates is halved, their area remaining the same, the capacitance is also doubled. These relationships give us a clue to the capacitance of two capacitors connected in parallel as shown in Figure 4.6a. The total area of the plates is effectively the sum of the two areas, so the total capacitance C of the combination is found by adding their values together.

$$C = C_1 + C_2 \text{ (parallel)}$$

Thus, if two capacitors of 10 µF and 50 µF are connected in parallel, the combined capacitance is 60 µF. However, if two

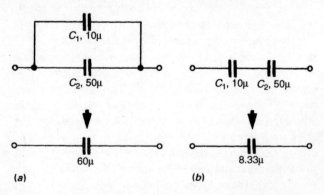

Figure 4.6 *Two capacitors connected in (a) parallel; and (b) series.*

capacitors are connected in series as shown in Figure 4.6b, the formula for finding their total capacitance is

$$C = \frac{C_1 \times C_2}{C_1 + C_2} \text{(series)}$$

The combined capacitance is equal to the product of their individual values divided by the sum of their individual values. Accordingly, if the above two capacitors of 10 μF and 50 μF are connected in series, the total capacitance of the combination is given by

$$C = \frac{10 \times 50}{10 + 50} = \frac{500}{60} = 8.33 \ \mu F$$

The combined capacitance of three capacitors connected in series is given by the equation

$$\frac{1}{C} = \frac{1}{C_1} + \frac{1}{C_2} + \frac{1}{C_3}$$

and so on. Note that when two or more capacitors are connected in parallel, their combined capacitance is more than the capacitance of the largest value. When they are connected in series, their combined capacitance is less than the smallest value capacitor. You should compare these equations with the equations for series and parallel resistor combinations in Chapter 3.

4.6 Charging and discharging capacitors

Many timers and oscillators are based on the simple circuit shown in Figure 4.7. Here a capacitor, C_1, and a resistor, R_1, are connected in series with a battery of e.m.f., E. On closing the

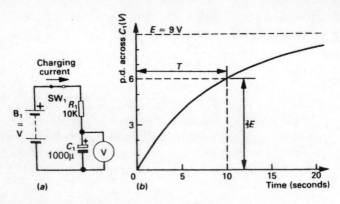

Figure 4.7 *The time constant of a RC combination.*

switch, SW_1, the p.d. across the capacitor rises as shown by the graph. The p.d. rises relatively fast at first and then more slowly as the p.d. approaches the e.m.f., E, of the battery. The delay in the charging (and discharging) of a capacitor is the key to understanding the use of capacitors in the design of timers and oscillators. The time taken for the p.d. across the capacitor to rise to two-thirds of E is known as the time constant of the capacitor/resistor circuit, or *RC* circuit. The time constant, T, is dependent on the values of both R_1 and C_1 and is given by the simple equation

$$T = R_1 \times C_1$$

This equation gives the time constant in seconds if C_1 has a value in farads and R_1 has a value in ohms. For example, suppose the two values are $C_1 = 1000$ μF and $R_1 = 10$ k . Thus $C_1 = 1000 \times 10^{-6}$ F and $R_1 = 10 \times 10^3 = 10^4$. Therefore, $T = 10^{-3}$ F $\times 10^4 = 10$ s. This means that if $E = 9$ V, it takes 10 seconds for the p.d. across the capacitor to rise from 0 V to two-thirds of E (6 V). In practical timer and oscillator circuits, the time constant is a convenient way of estimating the rate of

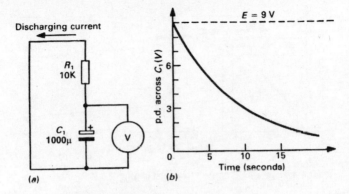

Figure 4.8 Discharging a capacitor.

charge of a capacitor. Note that, to be completely accurate, the p.d. should be measured rising to 63% of E. However, 'two-thirds' (67%) is accurate enough for most practical circuits since many electrolytic capacitors have a tolerance of +50% or more.

Just to complete our examination of the RC circuit, suppose the capacitor is discharged once it has been charged. Figure 4.8 shows what happens if the battery is isolated and the capacitor is allowed to discharge through resistor R_1. The graph shows that the p.d. across the capacitor falls fast at first and then more slowly as the voltage across it approaches 0 V. Incidentally, the graphs for the charge and discharge of a capacitor are known, mathematically, as exponential curves. We do not need to look at the mathematical equation for these graphs and how they help to define the time constant, but they do have an interesting property. This is shown in Figure 4.9 for the charging curve.

Suppose we carried on timing the charging of the capacitor after it had reached two-thirds of E (6 V). We should find that after another time constant of 10 s, the p.d. across the capacitor would have risen by two-thirds of the remaining p.d. Since the remaining p.d. is 3 V (9 V – 6 V), the p.d. would rise by a further 2 V to 8 V. After the next time constant of 10 s, the p.d. across

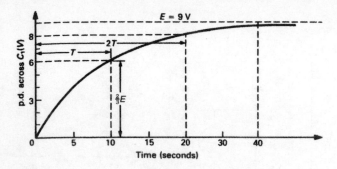

Figure 4.9 *The effects of successive time constants.*

the capacitor would have risen to two-thirds of 1 V (9 V – 8 V), i.e. by 0.67 V to 8.67 V. And so on. You can see that after three time constants (after 30 s in the example), the capacitor is almost fully charged, the p.d. across it having risen to 8.67 V, only 0.33 V short of its final p.d. Theoretically, the capacitor can never become fully charged (a characteristic of exponential curves), but after five or so time constants we can consider it to be fully charged.

4.7 The 555 timer monostable

Figure 4.10 shows the systems diagram of an electronic timer. The timer is 'triggered' by an input signal and produces an output signal that lasts for a time T seconds. Now in order to

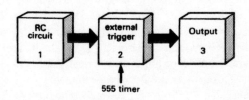

Figure 4.10 *The electronic system of a timer.*

make this function possible we need to connect together three black boxes as shown in Figure 4.10:

(a) *black box 1 is the RC combination discussed above;*
(b) *black box 2 is a device that detects when the p.d. across the capacitor has reached a certain value;*
(c) *black box 3 is an output circuit that indicates that timing is in progress.*

Now black box 2 must be a device that:

(a) *makes the output signal go HIGH when it receives a trigger signal; and*
(b) *returns the output signal to LOW when it detects that the p.d. across the capacitor in the RC combination has risen by a certain amount.*

Nowadays, a circuit designer would choose an integrated circuit (IC) to carry out the function of black box 2. And the most popular of the ICs available for this job is the device designated by manufacturers as the '555 timer' or 'triple-5 timer'. It is shown in Figure 4.11, and comprises a small black

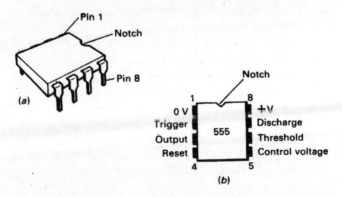

Figure 4.11 *The integrated circuit 555 timer: (a) what it looks like; and (b) the identity of its pins.*

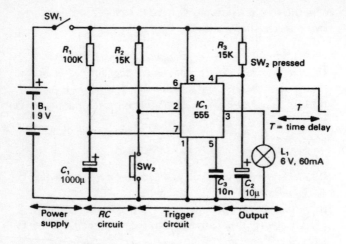

Figure 4.12 *A timer based on the 555 IC.*

plastic package having eight terminal pins for connecting it into a circuit. A silicon chip about 2 mm by 2 mm in area is inside the package.

Now we do not need to know anything about the workings of this chip to use it as black box 2 in the timer system. Figure 4.12 shows the circuit equivalent of the three interconnected black boxes. The main features of this circuit are as follows.

On closing switch SW_1 to switch on the power supply, the output signal is LOW, and the lamp L_1 is switched off. At this point, the capacitor C_1 is discharged (by a transistor on the 555 timer chip). Momentary pressing of the push-to-make, release-to-break switch, SW_2, triggers the timer. This opens the internal switch of the 555 timer and simultaneously makes the output of the 555 timer go HIGH. The lamp switches on and the capacitor C_1 begins to charge through resistor R_1. The lamp remains lit until the p.d. across the capacitor reaches two-thirds (exactly) of the e.m.f. of the supply. When the voltage across the capacitor reaches 6 V (two-thirds of 9 V), this is sensed by terminal pin 6 of the 555. At this instant, the internal switch on the 555 closes, C_1 is instantly discharged via pin

7, and the output of the 555 goes LOW. The lamp switches off and the circuit now waits for another trigger signal.

The time that the lamp remains lit is given by the equation $T = 1.1 \times C_1 \times R_1$. So, using the values in Figure 4.12,

$$T = 1.1 \times (1000 \times 10^{-6} \text{ F}) \times (100 \times 10^3 \text{ }\Omega)$$
$$= 110 \text{ s}$$

5

diodes and rectification

We know that the mains electricity supply is alternating current (ac) and that a direct current (dc) supply is needed for a battery charger for a digital camera, for example. But how does a.c. get to be d.c.? A semiconductor device called a diode will help us answer this question. First, we look at the basic properties of the diode, noting its ability to allow current to flow through it one way but to prevent current flowing through it in the opposite direction. Thus, in the context of this chapter, we shall refer to the diode more accurately as a rectifier, which distinguishes it from a light-emitting diode (LED), for example, which is designed to emit light when current flows through it in the easy flow direction. To make rectification clear, waveforms on a simulated oscilloscope screen are drawn to show the difference between half-wave rectification and full-wave rectification. Finally, we describe the use of a smoothing capacitor to make the output voltage ready, say, for charging a battery.

5.1 What diodes do

Rectification involves the use of one or more components called diodes or simply rectifiers. They behave as one-way 'valves' to control the current flowing through them. Indeed, the first diodes were thermionic valves (Chapter 1), so called because they allow electrons to pass easily through them in one direction only and not in the opposite direction. The symbol for a diode is shown in Figure 5.1a. Two types of diode are shown in Figure 5.1b, one that allows a maximum current of 1 A to flow through it and the other a maximum of 13 A. Note that the arrow of the diode symbol points in the direction of conventional current flow through it, so electrons flow through the diode in the direction opposite to this arrow. Conventional current flows through a diode from its anode terminal to its cathode terminal. On diodes encased in plastics, the cathode terminal is often marked with a red, black or white band. When the anode of a diode has a more positive voltage than the cathode, the diode is said to be forward biased and current flows easily through it (Figure 5.2a). When the anode is more negative than the cathode, it is said to be reverse biased and no current flows through it (Figure 5.2b).

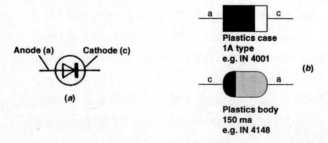

Figure 5.1 *Rectifier diodes: (a) general symbol; and (b) the appearance of two types.*

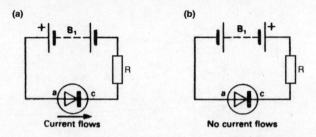

Figure 5.2 (a) A forward-biased diode; and (b) a reverse-biased diode.

5.2 Using diodes as rectifiers

An oscilloscope is used in the circuit of Figure 5.3 to see how a diode changes, or rectifies, a.c. to d.c. A 50 Hz alternating current supply of a few volts maximum value is connected across diode, D_1, which is also connected in series with resistor, R_1. The oscilloscope monitors the changes of voltage across resistor R_1. With SW_1 closed to short-circuit D_1, the oscilloscope displays an a.c. waveform, as might be expected since it is connected across the a.c. supply. When SW_1 is opened, the trace

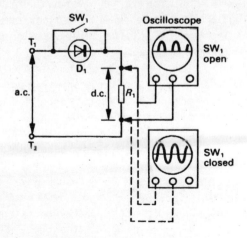

Figure 5.3 Half-wave rectification.

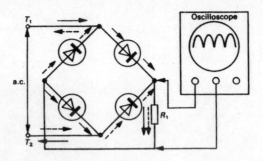

Figure 5.4 *Full-wave rectification*

shows that one half of the a.c. waveform is removed by the diode. This happens because D_1 is alternately forward-biased and reverse-biased by the a.c. supply. The diode allows current to flow through it and R_1 only when terminal T_1 of the supply is positive with respect to terminal T_2. The current through R_1 is a varying d.c. and the circuit is called a half-wave rectifier. Note that the half-wave rectifier 'loses' one half of the a.c. waveform.

A better rectifier circuit is shown in Figure 5.4; this is known as a full-wave rectifier. The 'lost' half wave now contributes to the d.c. output of the rectifier circuit since current flows through resistor R_1 in the same direction for both halves of the a.c. waveform. Thus, when terminal T_1 of the a.c. supply is positive with respect to T_2, current flows through R_1 in the direction shown by the full-line arrows. Moreover, when T_2 is positive with respect to T_1, the direction of the current through R_1 is unaltered and is represented by the dotted-line arrows.

The outputs of the half-wave and full-wave rectifier circuits are a varying d.c. voltage and they need to be smoothed to produce a steady d.c. supply. Smoothing is achieved with a capacitor, C_1 as shown in Figure 5.5 for the half-wave rectifier. When terminal T_1 is positive, the diode conducts and C_1 charges up to the peak voltage of the a.c. supply. During this time, current is also flowing through R_1. When the voltage at T_1 begins to fall during the first half cycle of the supply, current continues

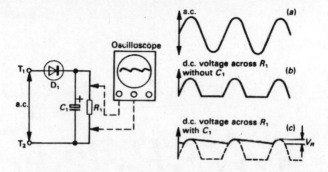

Figure 5.5 *Using a smoothing capacitor. The graphs show: (a) unrectified a.c.; (b) half-wave rectified d.c.; (c) smoothed a.c.*

to flow through R_1, and C_1 discharges through it. If the time constant, $T = C_1 \times R_1$ is long enough, C_1 discharges little during the second half cycle when D_1 is reverse-biased. The effect is a smoothed d.c. supply as shown by the graphs.

The d.c. output voltage across R_1 has a small ripple voltage superimposed on it caused by the small fall in voltage across C_1 during the second half cycle. For good smoothing action, the product $R_1 \times C_1$ must be larger than 1/50 second (0.02 s), which is the time between peak voltages across R_1 for a 50 Hz a.c. supply. For efficient smoothing of the d.c. output, the product $R_1 \times C_1$ ought to be about five times larger than 0.02 s. Thus if $R_1 = 1 \text{ k}\Omega$, $C_1 \times R_1 = 0.1$ s and, since $C_1 = T/R_1$

$$C = (0.1/R_1) = (0.1/10^3) = 10^{-4} \text{ F} = 100 \text{ }\mu\text{F}$$

In practical rectifier circuits, smoothing capacitors are of the electrolytic variety and have values between about 470 µF and 2200 µF. Perhaps you can see why a capacitor of lower value would suffice to smooth the varying d.c. voltage produced by a full-wave rectifier?

6

amplifiers and transistors

When John Bardeen, Walter Brattain and William Shockley invented the transisitor at the Bel Labs in 1947, no one really knew quite what to do with the invention. But once circuit designers realized that the tiny switches would enable products to be built smaller, more reliably and with less power consumption than with conventional electronic valves, the market started to develop. Five years later the first practical application appeared – a hearing aid. Now this was not a large market but it did demonstrate the key features of the new technology: it enabled something to be done that wasn't possible with the previous valve technology and it helped to enhance the quality of life for those in need. These two features, along with the industry's ability to reduce the production cost of a transistor by about 30 per cent a year, subsequently proved to be fundamental drivers behind today's enormously profitable electronics industry.

6.1 What amplifiers do

Figure 6.1 shows what an amplifier does. It increases the power of the signal passing through it. Thus a low-power signal, P_{in}, enters from the left, energy is drawn from the power supply, and a signal of higher power, P_{out}, leaves from the right. Note that an amplifier cannot increase the power of a signal without power being drawn from somewhere else, such as from a battery or some other source of electrical energy. In systems diagrams, it is usual to represent an amplifier black box as a triangle. In effect, this symbol is an arrowhead representing the direction in which the signal travels from the input to the output of the amplifier.

Now the ratio of the output power to the input power is the power gain, A, of an amplifier. Thus

$$\text{power gain} = \frac{\text{output power}}{\text{input power}}$$

$$\text{or } A = \frac{P_{out}}{P_{in}}$$

The power of electrical signals is measured in watts. Suppose the output power of an amplifier is equal to 50 W. If the input power is equal to 0.01 W (10 mW), the power gain of the amplifier is given by

$$A = \frac{50 \text{ W}}{10 \text{ mW}} = 5000$$

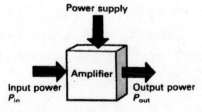

Figure 6.1 *The function of an amplifier.*

Thus, the output power from this amplifier is 5000 times more than the input power.

Amplifiers are very widely used in electronic systems, for example hi-fi amplifiers increase the power of audio frequency signals from compact discs before delivery to a loudspeaker. So let us take a closer look at the audio amplifier, which is perhaps the one you are most likely to come across.

6.2 Audio amplifiers

Audio amplifiers are to be found in hi-fi amplifiers, car radios, mobile phones, cassette players, MiniDisk players, hearing aids and satellite communications systems, to name but a few. In these systems, their purpose is to increase the power of audio frequency (AF) signals in the range of human hearing, which is between 20 Hz to 20,000 Hz (20 kHz). A 20 Hz audio frequency signal sounds like a 'buzz', and one of 10 kHz sounds like a high-pitched whistle. Few adults can hear AF signals above about 16 kHz, though many young children and animals do hear sounds that have frequencies higher than this. As we get older, our ears tend to become increasingly less sensitive to audio frequencies over 10 kHz.

The easy way to amplify a weak audio signal is to use an integrated circuit designed for the job. A good example of such an audio IC is the TBA820M device shown in Figure 6.2. The TBA820M is housed in an 8-pin dual-in-line package that protects the 2 mm by 2 mm square silicon chip on which the amplifying circuit is made. When it is operated from a 12 V power supply, this device provides a maximum audio power of 2 W from an 8 ohm loudspeaker. It is actually designed to operate from any power supply voltage in the range 3 V to 16 V, though at 3 V its power output is much less than at 16 V. The manufacturers of this audio amplifier have also made sure that when it is not actually amplifying a signal it draws very little current from the power supply – it is said to have a low quiescent current.

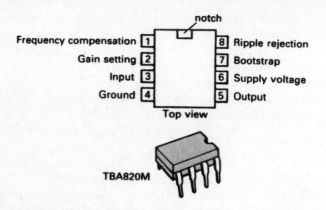

Figure 6.2 *The TBA820M audio amplifier.*

6.3 The bandwidth of an amplifier

Most audio amplifiers are designed to amplify all frequencies within a specified band of frequencies by the same amount. This is explained by means of the frequency response graph shown in Figure 6.3. The level part of the graph is typical of the 'flat' frequency response of a hi-fi amplifier. However, the power gain of most audio amplifiers falls off sharply at frequencies below about 20 Hz and above about 20 kHz. Surprisingly, the power gain of some good-quality audio

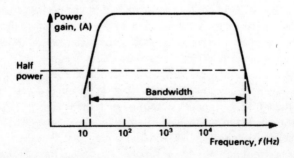

Figure 6.3 *The frequency response graph of an audio amplifier.*

amplifiers only begins to fall off at frequencies above 40 kHz. Manufacturers claim that this ability to amplify frequencies well above the range of human hearing improves the sound reproduction within the audio frequency range.

The frequency range over which the power gain does not fall below half the maximum power gain is called the bandwidth of the amplifier. Note that frequencies are plotted 'logarithmically' on the frequency response graph. This means that ten-fold increases of frequency occupy equal distances along the horizontal axis. Thus, there is equal spacing between 10 Hz and 100 Hz, and between 100 Hz and 1000 Hz, and so on, so that the graph can accommodate the wide range of frequencies and show clearly the amplifier's frequency response in the 20 Hz to 20 kHz frequency range. If the calibrations along the frequency axis were linear, this 100 fold frequency range would occupy only about one-tenth of the full bandwidth of 20 kHz. As it is, it occupies about one-half of the 20 kHz range.

Though a good-quality amplifier has a flat frequency response over a wide range, it is common for manufacturers to provide a means to adjust this response to suit individual tastes. This may take the form of a single potentiometer, a 'tone control' that can be adjusted to select relatively more high or low audio frequencies. Or the device might be equipped with a 'graphic equalizer', which allows control of the frequency response in the low, middle and upper frequency range of the audio band.

6.4 Types of transistor

Invented in 1947 by the three-man team of Bardeen, Shockley and Brattain at the Bell Telephone Laboratories, USA, the transistor became the most important basic building block of almost all circuits. Transistors were first used singly in circuits such as in the early 'transistor' radios of the late 1950s but, by the early 1970s, silicon chips comprising several hundred transistors were being made. Nowadays, the most complex integrated circuits contain upwards of a million transistors.

Transistors are still used as discrete, i.e. individual, components in circuits. Figure 6.4 shows the distinguishing

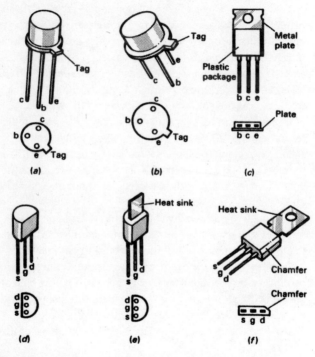

Figure 6.4 *A selection of bipolar and unipolar transistors.*
In TO18 outline, (a) npn types BC108 and 2N2222; and pnp types BC478 and 2N2907;
in TO39 outline, (b) npn type 2N3019 and pnp type 2N2905;
in TO 220 outline, (c) npn, type TIP31A and pnp type TIP32A;
in TO92(D) outline, (d) n-channel junction gate FET, type BF245C;
in TO237 outline, (e) n-channel metal-oxide field-effect transistor (MOSFET), type VN10LM;
in TO202(B) outline, (f) n-channel MOSFET, type VN46AF.

features of a small selection of transistors that fall into two main categories depending on the way n-type and p-type semiconductors are used to make them. One sort is the bipolar transistor (the sort invented in 1947) of which there are two types, npn and pnp, as shown by the examples in Figure 6.4a to c. The second sort is the unipolar transistor (a later invention) that is also called the field-effect transistor (FET). Examples of the FETs, of which there are two sorts, n-channel and p-channel, are shown by the examples in Figure 6.4d to f. Bipolar transistors have three leads called emitter (e), base (b) and collector (c); and the leads of an FET are source (s), gate (g) and drain (d). These transistors are housed in a metal or plastic package. In many of the 'metal can' types, e.g. the TO18 shape, the collector lead is also connected to the metal can. Transistors used for high power applications have a flat metal side to them, e.g. the TO220 shape, to which a heat sink can be bolted to help the transistor dissipate excess heat produced within it. The symbols for bipolar and unipolar transistors are shown in Figure 6.5.

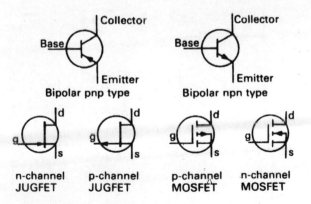

Figure 6.5 *Transistor circuit symbols.*

6.5 Transistors as electronic switches

Despite the advantages of using integrated circuits in circuit design, millions of individual transistors are made each year, some of which are used in simple control circuits like the one shown in Figure 6.6. This circuit switches on the lamp L_1 when the light-dependent resistor LDR_1 is covered as it is a dark-operated switch. Similar circuits are used in automatic streetlights that come on at dusk. This electronic switch comprises three building blocks: a sensor based on resistors R_1 and LDR_1 that make up a potential divider; a current amplifier based on transistor Tr_1, and an output device, in this case the lamp L_1. The voltage at point X controls the on/off action of this circuit. This voltage rises when LDR_1 is covered, and falls when LDR_1 is illuminated. When the voltage at X rises, the current through the lamp increases; when it falls the lamp current decreases.

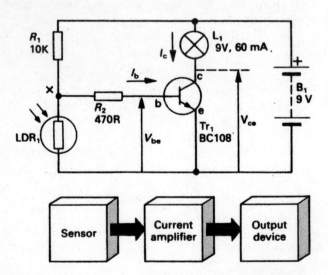

Figure 6.6 *Single transistor switching circuit comprising three building blocks.*

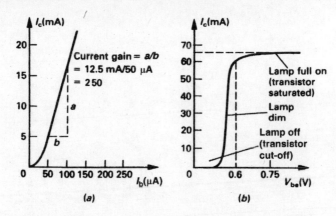

Figure 6.7 *Characteristics of an npn transistor: (a) variation of collector current with base current; (b) variation of collector current with base voltage.*

Figure 6.7a shows the action of the npn transistor as a current amplifier; small changes of base current produce large changes in the collector current. The ratio I_c/I_b is the current gain, 250 in this case. But I_b and hence I_c is controlled by the base–emitter voltage, V_{be}, as shown in Figure 6.7b, a graph known as a transfer characteristic. If V_{be} is less than 0.6 V, I_b and I_c are both zero, and the lamp is off. In this state, the collector–emitter resistance is very high and the collector–emitter voltage V_{ce}, is +9 V. The transistor is said to be 'cut off' and the transistor switch is 'open'. As V_{be} increases above 0.6 V, I_b increases, and the amplifying action of the transistor produces a much larger collector current, I_c. The lamp brightens until V_{be} has reached about 0.75 V. In this state, the collector–emitter resistance is so low that V_{ce} is nearly zero. The transistor is then said to be saturated since any further increase in V_{be} does not increase I_c. The transistor switch is 'closed'.

7

logic
gates and
Boolean
algebra

We have already met the idea of a digital circuit in Chapter 2, and at the end of the previous chapter we saw how the transistor can be used as an electronic switch. These switches have input and output signals that have two values, either on or off (1 and 0). By way of introduction to this chapter, we will sketch the work of George Boole, born in Lincoln, England, in 1815. He was lucky enough to have a father who passed on his own love of mathematics, and this passion led him to use maths as a solution to mathematical problems. As he delved deeper into his own work he was determined to find a way to encode logical arguments into a language that could be manipulated and solved mathematically. He came up with a type of linguistic algebra – which subsequently became known as Boolean algebra – the three most basic functions of which were (and still are) AND, OR and NOT.

7.1 What logic gates do

The basic building blocks that make up all digital systems are simple little circuits called logic gates. A logic gate is a decision-making building block that has one output and two or more inputs as shown in Figure 7.1. Two examples of these decision-making logic gates were introduced in Chapter 3. These were the AND gate and the OR gate using mechanical switches. In this chapter, logic gates are contained within integrated circuits that operate electronically instead of mechanically. The input and output signals of these gates can have either of two binary values, 1 or 0. The binary value of the output of a gate is decided by the values of its inputs, and a truth table for a logic gate shows the value of the output for all possible values of the inputs. Like the other gates below, AND and OR gates are known as combinational logic gates. This term arises because their outputs are the logical (i.e. predictable) result of a particular combination of input states.

Logic gates are used in many types of computer, control and communications systems, and especially in calculators and digital watches, and other devices that have digital displays. Figure 7.2 shows the usual place of logic gates in digital systems. They have the intermediate task of receiving signals from sensors such as keyboard switches and temperature sensors, making decisions based on the information received, and sending an output signal to a circuit that provides some action, such as switching on a motor, activating a 'go' or 'stop' lamp. Although logic gates can be designed using individual

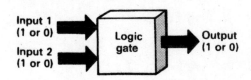

Figure 7.1 *A logic gate has two or more inputs and one output.*

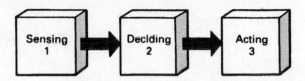

Figure 7.2 *The function of a digital system: logic gates are used in building block 2.*

diodes and transistors, most digital circuits now make use of logic gates in integrated circuit packages. There are two main 'families' of these digital ICs; one is called transistor–transistor logic (TTL), and the other complementary metal-oxide semiconductor logic (CMOS).

7.2 Symbols and truth tables of logic gates

Two alternative systems are in use for showing the symbols of logic gates in circuit diagrams, the American 'Mil Spec' system and the British Standards system. Figure 7.3 summarizes these symbols for six logic gates, the AND, OR, NOT, NAND, NOR and exclusive-OR gates. The American Mil Spec symbols are widely preferred, since their different shapes are easily recognized in complex circuit diagrams although in the UK there is some pressure to adopt the British Standards symbols.

The AND gate gives an output of logic 1 when all inputs are at logic 1, and an output of logic 0 if any or all inputs are at logic 0. Therefore, an AND gate is sometimes called an 'all-or-nothing gate'. For the two-input AND gate shown in Figure 7.3a, the output, S, is at logic 1 only when input A and input B are at logic 1. The truth table for the two-input AND gate gives the state of the output, S, for all combinations of input states – hence the term 'combinational logic' is used to describe logic systems using gates like the AND gate.

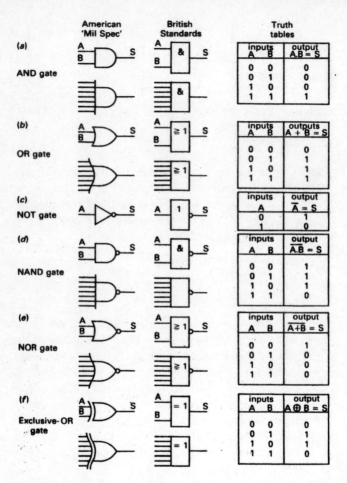

	American 'Mil Spec'	British Standards	Truth tables		

(a) AND gate

inputs A	B	output A.B = S
0	0	0
0	1	0
1	0	0
1	1	1

(b) OR gate

inputs A	B	outputs A + B = S
0	0	0
0	1	1
1	0	1
1	1	1

(c) NOT gate

inputs A	output $\overline{A}$ = S
0	1
1	0

(d) NAND gate

inputs A	B	output $\overline{A.B}$ = S
0	0	1
0	1	1
1	0	1
1	1	0

(e) NOR gate

inputs A	B	output $\overline{A+B}$ = S
0	0	1
0	1	0
1	0	0
1	1	0

(f) Exclusive-OR gate

inputs A	B	output $A \oplus B$ = S
0	0	0
0	1	1
1	0	1
1	1	0

Figure 7.3 *Symbols and truth tables for logic gates.*

For the two-input OR gate shown in Figure 7.3b, the output, S, is at logic 1 when either input A or input B, or both inputs, are at logic 1. Thus, the OR gate is sometimes called an 'any-or-all' gate. The truth table for the two-input OR gate gives the state of the output, S, for all combinations of input states.

For the NOT gate shown in Figure 7.3c, the output, S, is simply the inverse of the input A. Thus, if the state of the input is logic 1, the output state is logic 0 and vice versa. For this reason the NOT gate is also called an inverter. The truth table for the NOT gate is simple – it has only two lines.

The NAND (or NOT-AND) gate gives an output that is the inverse of the AND gate. Thus, for the two-input NAND gate shown in Figure 7.3d, the output, S, is at logic 0 when both input A and input B are at logic 1. The truth table for the two-input NAND gate gives the state of the output, S, for all combinations of input states. Compare it with the truth table of the AND gate.

The NOR (or NOT-OR) gate gives an output that is the inverse of the OR gate. Thus for the two-input NOR gate shown in Figure 7.3e, the output, S, is at logic 0 when either input A or input B, or both inputs, are at logic 1. The truth table for the two input NOR gate gives the state of the output, S, for all combinations of input states.

The two-input Exclusive-OR (or XOR) gate shown in Figure 7.3f does something that the OR gate of Figure 7.3b does not: it is a true OR gate for it only gives an output of logic 1 when either, but not both, of its inputs is at logic 1. The truth table summarizes the action of the XOR gate and you should compare it with that for the OR gate.

7.3 Introducing Boolean algebra

In 1847, George Boole invented a shorthand method of writing down combinations of logical statements that are either 'true' or 'false'. Boole proved that the binary or two-valued nature of his 'logic' is valid for symbols and letters as well as for words. His mathematical analysis of logic statements became known as Boolean algebra.

Since digital circuits deal with signals that have two values, Boolean algebra is an ideal method for analysing and predicting the behaviour of these circuits. For example, 'true' can be regarded as logic 1, i.e. an 'on' signal, and 'false' as logic 0, i.e.

an 'off' signal. However, Boolean algebra did not begin to have an impact on digital circuits until 1938 when Shannon applied Boole's ideas to the design of telephone switching circuits.

The table below shows how Boolean algebra represents the functions of the six logic gates, AND, OR, NOT, NAND, NOR and XOR, discussed above.

Logic statement	Boolean expression
(a) AND gate Input A and input B = output S	$A . B = S$
(b) OR gate Input A or input B = output	$A + B = S$
(c) NOT gate Not input A = output S	$\overline{A} = S$
(d) NAND gate Not input A and input B = output S	$\overline{A} . \overline{B} = S$
(e) NOR gate Not input A or input B = output S	$\overline{A + B} = S$
(f) Exclusive-OR gate Input A or input B = output S (excluding input A and input B)	$A \oplus B = S$

Note the use of the following conventions in Boolean algebra:

(a) *two symbols are ANDed if there is a 'dot' between them;*
(b) *two symbols are 'ORed' if there is a '+' between them;*
(c) *a bar across the top of a symbol means the value of the symbol is inverted.*

The use of the bar across the top of a symbol is very important in Boolean algebra. Since A and B can take values of either 0 or 1, $\overline{1} = 0$ and $\overline{0} = 1$ as in the NOT gate. If the bar is used twice, this represents a double inversion. Thus, $\overline{\overline{1}} = 1$, and $\overline{\overline{0}} = 0$. What do you think happens to the value of the symbol if the bar is used three times?

8

op amps
and
control
systems

Operational amplifiers (op amps) are at the heart of control systems. They are used in integrated circuit form usually to compare an input signal with a desired output signal, involving making adjustments to achieve a preset output such as temperature, position or speed. To achieve this type of control we shall look at the idea of positive and negative feedback. We shall then turn our attention to the connections needed for an operational amplifier to be used as a comparator in control systems. We shall be looking at the op amp as a black box and will not be concerned with the details of its inner working. This is normal practice when designing any electronic system whereby we are concerned only with what the black box does. We have already met this idea in Chapter 4 when using the 555 timer.

8.1 The basic features of control systems

Control engineering is a vast field and ranges from simple thermostatic control systems for, say, the control of temperature in a tropical fish tank, to advanced position-control systems aboard spacecraft exploring the Solar System. But whatever their level of sophistication, all control systems have certain common features. The simplest form of control is open-loop control. Figure 8.1a shows its three basic elements. Building block 1 called 'desired output' is what the user of the control system wants as the 'actual output' shown by building block 3. Building block 2, the 'controller', makes the output possible after the input has been set. A typical example of an open-loop control system is a domestic light dimmer switch. The desired light level is selected by the amount a control knob is turned.

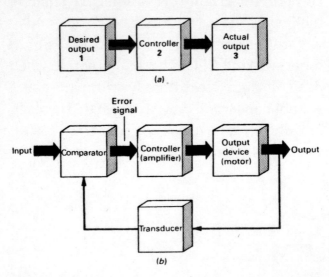

Figure 8.1 *Two basic types of control system: (a) open-loop control; (b) closed-loop control.*

However, if the light dims because of a partial power failure, there is no feedback between the output (the amount of light produced) and the input (the control setting) to enable it to make changes to the actual output once the input has been set.

Of course, having set the dimmer switch you could intervene to bring the light level back to what you wanted. Now you have taken action and provided feedback, the dimmer switch has become a closed-loop control system. But the system is a closed-loop control system only as long as you, the human operator, remain on duty. In automatic closed-loop control systems, a monitoring device replaces the action of the human operator, and the automatic device can't go to sleep! Figure 8.1b shows the basic elements of a closed-loop control system. The system includes a transducer for monitoring the prevailing state of the actual output and converting it into a form similar to the signal representing the desired output. These two signals are compared to produce a difference, or error, signal that is then used to control the system. Thus, closed-loop control systems are 'error-actuated'. Modern control systems use a variety of electrical transducers for producing the feedback signal. For example, temperature can be monitored with a thermistor, position with a variable resistor, e.g. a potentiometer, and forces with a strain gauge.

8.2 The design of a simple thermostat

The simple thermostat circuit shown in Figure 8.2 is an example of a closed-loop control system. It comprises three parts: a thermistor temperature sensor, Th_1, which is one component of a voltage divider; a comparator based on an integrated circuit, IC_1, called an operational amplifier; and a single transistor amplifier, Tr_1, which opens and closes the relay contacts to control the power to the lamp. The detailed operation of an operational amplifier (op amp) is described below. The op amp shown in Figure 8.2 compares the voltage set on pin 3 with that on pin 2. It produces a positive voltage to

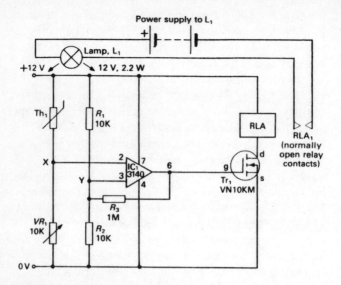

Figure 8.2 *Simple thermostat circuit.*

switch the transistor on when the voltage on pin 2 is less than that on pin 3.

In this simple circuit, a 12 V filament lamp produces the heat. In use, the lamp could heat the air in a small box that houses tropical insects, for example. The thermistor senses the temperature in the box. Suppose VR_1 is set so that the lamp just switches on so that the output voltage from the comparator is positive. In this case, the voltage on pin 2 of the comparator is slightly lower than the voltage on pin 3. As the heat from the lamp warms the air inside the box, the resistance of the thermistor falls. When this fall makes the voltage at pin 2 rise above that on pin 3, the output voltage from the comparator falls to 0 V and the heater is switched off. Thus, the temperature of the box rises or falls to a value determined by the reference voltage set on pin 3 by VR_1. The op amp provides the active sensing component that acts to 'close the loop'. A scale could be fitted to the spindle of VR_1 so that the thermostat could be used to stabilize preset temperatures once the thermostat had been calibrated.

8.3 The basic properties of op amps

The principal electronic component in the thermostat and servosystem described in the preceding two sections is an integrated circuit operational amplifier or op amp. These two designs use an op amp to compare a reference voltage set on one of its two inputs with a varying voltage on its other input. The output voltage of the op amp changes abruptly when the varying voltage is slightly higher or lower than the reference voltage. This abrupt change in output voltage is used to switch a relay on or off to form the basis of a simple temperature control system or a position control system. The op amp is such an important integrated circuit in control systems and instrumentation systems that an explanation is needed about what it does, and what is particularly useful about its characteristics.

The appearance of one type of op amp is shown in Figure 8.3a. This is an 8-pin dual-in-line (d.i.l.) package. Figure 8.3b shows how the pins are numbered looking at the package from the top. Figure 8.3c shows the general connections to the op amp. Note its circuit symbol is a triangle, an arrowhead that shows the direction in which signal processing takes place. The op amp has two inputs (pins 2 and 3), one output (pin 6) and two power supply connections (pins 4 and 7). An op amp can be operated with one or two power supplies as shown in Figure 8.5. If it uses a single power supply, pin 4 is connected to 0 V and pin 7 to +V. Input and output signals are measured with respect to 0 V. With two power supplies, pin 4 is connected to the −V terminal of one supply, pin 7 is connected to the +V of the other supply, and the common connection of the two supplies provides the 0 V supply line. Now input and output voltages are measured with respect to this common connection.

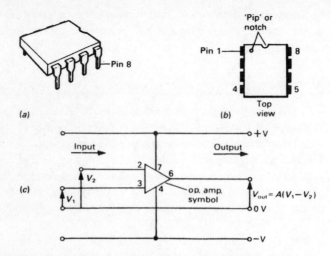

Figure 8.3 *The dual-in-line (d.i.l.) op amp package: (a) its general appearance; (b) its pin identification; (c) its circuit symbol.*

However, why does the op amp have two inputs? Figure 8.3c shows that one input (pin 2) has an input voltage of V_2 on it, and the second input (pin 3) has an input voltage of V_1 on it. What does the op amp do with these two input voltages? It amplifies the difference between them so that the output signal V_{out} is given by the following 'op amp equation':

$$V_{out} = A(V_1 - V_2) = A \times V_{in}$$

where $(V_1 - V_2)$ is the input voltage, V_{in}, and A is known as the voltage gain of the op amp. A is the number of times the output voltage is greater than the input voltage, and it is very high. Most modern op amps have voltage gains in excess of 100,000. Note that an op amp is not sensitive to the actual values of the voltages on its two inputs – it only 'sees' the difference between them.

One of the op amp's two inputs (pin 2) is called the inverting input and the other (pin 3) the non-inverting input.

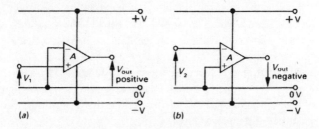

Figure 8.4 *Power supply connections to the 8-pin d.i.l. version of an op amp.*

What do these names mean? Figure 8.5 shows an op amp wired up with two power supplies, but with one of the inputs connected to 0 V. The 'op amp equation' shown above can be used to find out what the output voltage is in these two cases. If a positive input voltage, V_1, is applied to the non-inverting input (pin 3) and the inverting input is connected to 0 V, the output voltage, V_{out}, is positive and equal to $A \times (V_1 - 0) = A \times V_1$. However, if a positive input voltage, V_2, is applied to the inverting input (pin 2), the output voltage is given by $A \times (0 - V_2)$. That is, $-AV_2$. Thus, any difference between the two input voltages that makes the non-inverting input voltage more positive than the inverting input provides an amplified positive (above 0 V, and therefore 'non-inverted') output voltage. Moreover, any difference between the input voltages that makes

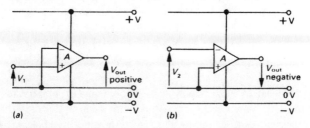

Figure 8.5 *The action of (a) the non-inverting input; and (b) the inverting input of an op amp.*

8 op amps and control systems

the inverting input voltage more positive than the non-inverting input voltage provides an amplified negative (below 0 V and therefore 'inverted') output voltage. It is common to indicate the inverting input of an op amp by a '−' sign to indicate its ability to invert the sign of an input signal, and the non-inverting input with a '+' sign to indicate its ability not to invert the sign of an input signal. You should not confuse these signs with the polarities of the power supplies.

Thus, we have two basic characteristics of an op amp:

1 *An op amp has a very high voltage gain, e.g. the 3140 op amp used in the thermostat of Figure 8.2 has a voltage gain in excess of 100,000. Hence op amps are sometimes referred to as 'packages of gain'.*
2 *An op amp responds to the difference of voltage between its two input terminals. Hence an op amp is sometimes called a difference amplifier.*

8.4 Using op amps as comparators

A comparator is a circuit building block that compares the strength of two signals and provides an output signal when one signal is bigger than the other. Clearly, the op amp is one type of comparator since its two inputs are able to compare the magnitude of two voltages as shown in Figure 8.6. It simply compares a voltage, V_{in}, here shown applied to the inverting input of the op amp, with a reference voltage, V_{ref}, applied to the non-inverting input. Thus any slight difference in voltage, $e = V_{ref} - V_{in}$, causes the output voltage to saturate. Since the op amp is operated from a dual power supply, the two possible saturation voltages are $+V_{sat}$ or $-V_{sat}$. If a single power supply were used, the saturation voltages would be $+V_{sat}$ and 0 V. To 'saturate' means to go to the maximum value. In this case, the maximum output voltage is limited to a volt or two below the supply voltage, e.g. 10 V for a 12 V power supply. The output voltage cannot take on any intermediate value between the

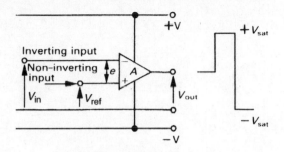

Figure 8.6 *An op amp used as a comparator.*

upper and lower saturation voltages since the gain of the op amp is so high. Thus if its gain is 100,000, the difference e between the two input voltages that makes the output voltage saturate at 10 V is (10 V/100,000) or 0.0001 V. If the difference exceeds this value, the output voltage remains saturated at 10 V.

In the design of the thermostat in Figure 8.3, the reference voltage is applied to pin 3 by two equal-value resistors R_1 and R_2. The voltage divider action of these two resistors sets a reference voltage of about 6 V on the non-inverting input, pin 3. This voltage is compared with the changing voltage on the inverting input, pin 2, determined by the resistance of the thermistor. Now suppose the variable resistor VR_1 is set so that the voltage on pin 2 is lower than that on pin 3, say 5.8 V compared with 6 V on pin 3. This makes the output voltage at pin 6 rise to the upper saturation voltage, about 10 V, and transistor Tr_1 switches on. Thus the relay is energized and power is supplied to the heater. The thermistor senses this rise in temperature and its resistance falls as explained in Chapter 3. The voltage on pin 2 therefore rises. As soon as this voltage exceeds 6 V, the output voltage falls to the lower saturation voltage, 0 V in this case. The transistor switches off and power is no longer supplied to the heater. As the temperature of the thermistor rises and falls, the output voltage of the comparator falls to 0 V and rises to about 10 V respectively.

op amps and
instrumentation
systems

Operational amplifiers are one of the most widely used integrated circuits today. They are used in numerous consumer, industrial and scientific devices. Their costs range from a few pence for well-known and popular op amps to more than £100 for specialist devices such as instrumentation amplifiers. Op amps usually have two inputs and a single output and may be packaged singly in an 8-pin or 14-pin dual-in-line package. As explained in this chapter, high input resistance and high voltage gain are important characteristics. They can be operated with either negative or positive feedback. Negative feedback enables the designer to tailor the intrinsically high gain to a defined lower gain determined solely by external resistor values. Positive feedback, on the other hand, provides improved switching function.

9.1 Electronics and measurement

Our senses are good at detecting changes in quantities like temperature, frequency and light intensity, but instruments are needed to give us values on which we can all agree. An instrument is not only able to provide the actual value of a quantity, but it can take readings in inaccessible places, such as in the middle of a grain store and on the surface of Mars. Furthermore, instruments can measure quantities such as atomic radiation, radio frequencies and atmospheric pressure, to which our senses are quite oblivious. Moreover, digital signals propagate more efficiently than analogue signals, largely because digital signals, that are well defined and orderly, are easier for electronic circuits to distinguish from noise that is chaotic.

Electronic instruments are common, and many of these are essentially digital. Digital watches, clocks, thermometers and weighing machines are electronic. So, too, are many of the instruments used in weather forecasting, on car instrument panels and in the flight decks of airliners. The systems diagram shown in Figure 9.1 summarizes the general function of these instruments. An instrumentation system comprises three basic building blocks: sensor, signal processor and a digital or analogue display.

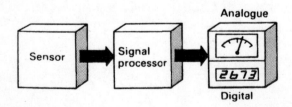

Figure 9.1 *The three main building blocks of an instrumentation system.*

9.2 Designing voltage amplifiers with op amps

Op amps are also used widely in the design of instruments. For example, an electronic thermometer uses an op amp to amplify the small e.m.f. that is generated by a thermocouple temperature sensor. In this application, the intrinsically high voltage gain of an op amp, that is so useful in the design of control systems, has been 'tamed' to provide a much reduced but accurately known voltage gain. How is this possible?

The technique used to control the voltage gain of an op amp is called negative feedback. There are two basic voltage amplifier circuits making use of negative feedback: the inverting negative feedback voltage amplifier (Figure 9.2a); and the non-inverting negative feedback voltage amplifier (Figure 9.2b). In these circuits the ratio of the output voltage V_{out} to the input voltage V_{in} is the voltage gain A of the amplifier. The equations governing this voltage gain are remarkably simple as the following table shows.

Amplifier circuit	Voltage gain, A =
Inverting negative feedback voltage amplifier	$V_{out}/V_{in} = -R_2/R_1$
non-inverting negative feedback voltage amplifier	$V_{out}/V_{in} = 1 + R_2/R_1$

Note two things about these equations:

1 *First, neither equation makes reference to the intrinsically high gain of the op amp; the gain in each case is determined only by the values of the external resistors – a surprising result, perhaps? You might be tempted to suggest removing the op amp from the circuits leaving only the two resistors!*

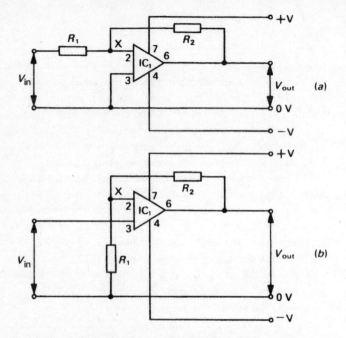

Figure 9.2 *The use of op amps as negative feedback voltage amplifiers:*
(a) an inverting amplifier; and (b) a non-inverting amplifier.

2 *Second, the negative sign in the equation for the inverting*
voltage amplifier means that the input voltage is inverted,
i.e. a positive input voltage produces a negative output
voltage, and vice versa. The non-inverting voltage amplifier
does not change the sign of the input voltage.

Thus supposing you choose resistor values of $R_2 = 100$ kΩ
and $R_1 = 10$ kΩ. In the case of the inverting amplifier the voltage
gain is −100 kΩ/10 kΩ = −10 times. And in the case of the
non-inverting amplifier, the voltage gain is 1 + 100 kΩ/10 kΩ =
11 times. An input voltage of +1 V produces an output voltage

of −10 V in the inverting amplifier, and of +11 V in the non-inverting amplifier. Here we are assuming that the op amps are operated from a dual power supply so that the output voltage is negative, i.e. below 0 V. Of course, any other values of resistors R_2 and R_1 could be used to obtain the voltage gain required for a particular application.

How do the circuits shown in Figure 9.2 manage to control the intrinsically high gain of the op amp, from values of 100,000 or more (known as the open-loop voltage gain, A_{vol}), to less than 1000 (known as the closed-loop voltage gain, A_{vcl})? To explain why, we need to assume that op amps have two ideal characteristics:

1 *They have an infinitely large open-loop voltage gain, i.e. A_{vol} infinity (actual op amps have open-loop gains in excess of 100,000)*
2 *They draw no current whatsoever from the source of the signals at either of their two inputs, i.e. I → zero (actual op amps have input currents less than 10^{-6} A).*

These ideal characteristics enable us to prove the two closed-loop gain equations. First, let us start with the inverting voltage amplifier shown in Figure 9.2, in which the input voltage, V_{in}, is applied to the inverting input of the op amp via resistor R_1. Resistor R_2 is a 'feedback resistor' that 'closes the loop' between the output (pin 6) and the inverting input (pin 2). The input side of R_1 is at a voltage of V_{in}, and the output side of R_2 is at a voltage of V_{out}, both voltages measured with respect to 0 V. So what is the voltage, V_x, at the join between the two resistors, i.e. at point X, the inverting input of the op amp? Figure 9.3a shows the connections at the ends of the resistors.

Note the connection to 0 V of the non-inverting input of the op amp. Now if we have a perfect op amp (that is one that has an infinitely high open-loop gain), there is no difference between the two input voltages: if pin 3 is at 0 V, pin 2 must

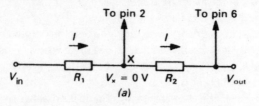

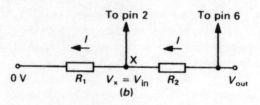

Figure 9.3 *Proving the closed-loop gain equations for (a) the inverting amplifier; and (b) the non-inverting amplifier.*

effectively be at 0 V. Of course, for practical amplifiers like the 741 and 3140, a very small difference of voltage (less than a microvolt) will exist between the two inputs. Thus for the perfect op amp, the voltage V_x at point X in Figure 9.3a is 0 V. This point is not actually connected to 0 V but it might just as well be. It is therefore called the virtual earth in the op amp circuit.

If we assume that the operational amplifier does not require any input current, we can concentrate our attention on working out the relationship between V_{out} and V_{in}. Let us assume that V_{in} is positive so that a current I flows through R_1 towards X. This same current flows through R_2 which is in series with it, since no current flows into pin 2. Thus, the following equations follow from the relationship between V, I and R:

> **For resistor R_1:** $I = (V_{in} - 0)/R_1$
> **For resistor R_2:** $I = (0 - V_{out})/R_2$

Note that the voltage difference across R_2 is $0 - V_{out}$ since current flows from pin 2 to pin 6. These equations give $V_{in}/R_1 = -V_{out}/R_2$ that can be rearranged to give

$$(V_{out}/V_{in}) = -(R_2/R_1) = A_{vcl}$$

This is the equation that was written down above.

Now let us look at the non-inverting amplifier shown in Figure 9.3b. In this case, the input voltage, V_{in}, is applied direct to the non-inverting input, pin 3, and R_1 is connected from pin 2 to 0 V. Note that R_1 and R_2 are connected in series as shown in Figure 9.5b. The voltage at one end of R_2 is V_{out}, and at one end of R_1 is 0 V. Again, before we can use the relationship between V, I and R to prove the gain equation, we have to know the voltage, V_x, at point X between the two resistors. Since the op amp is 'perfect', there is no difference of voltage between the two inputs. So as V_{in}, changes, V_x must follow these changes; $V_x = V_{in}$. In this case, note that neither pin 2 nor pin 3 is at 0 V, virtual earth. Let us assume that a current I flows through R_1. This same current (remember there is no current into pin 2) then flows through R_2 which is in series with it. Thus, the following equations can be written down.

> **For resistor R_1:** $I = (V_{in} - 0)/R_1$
> **For resistor R_2:** $I = (V_{out} - V_{in})/R_2$

These equations give $V_{in}/R_1 = (V_{out} - V_{in})/R_2$. Dividing through by V_{in} and rearranging gives

$$(V_{out}/V_{in}) = 1 + (R_2/R_1) = A_{vcl}$$

This is the equation written down above.

Thus, the closed-loop voltage gain of negative feedback amplifiers is independent of variations in the characteristics of the resistors, transistors and other components making up the integrated circuit inside the op amp. It does not matter whether we use an op amp having a gain of 100,000, or 1 million, or 250,000, the closed-loop gain is determined solely by the values of resistors connected externally to the op amp. The feeding back of part of the output voltage to the inverting input is known as negative feedback and its effect is to stabilize the voltage gain of the op amp to a value determined by the values of the external resistors. Thus, if the output voltage tends to rise, a small fraction of this rise is applied to pin 2. Since this is the inverting input of the op amp, the op amp acts to lower the output voltage. Likewise, the op amp counterbalances a decreasing output voltage by trying to raise the output voltage. The only stable value of the output voltage is that determined by the values of the two external resistors.